Sound

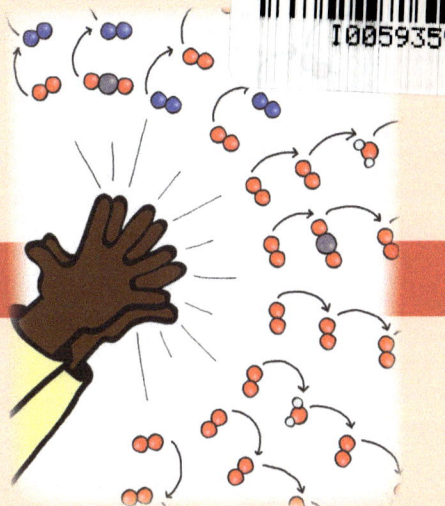

Illustrations: Janet Moneymaker
Design/Editing: Marjie Bassler

Sound
ISBN 978-1-950415-27-4

Published by Gravitas Publications Inc.
Imprint: Real Science-4-Kids
www.gravitaspublications.com
www.realscience4kids.com

What happens when
you clap your hands?

Sound!

Your hands push molecules in the air, making the air molecules wiggle.

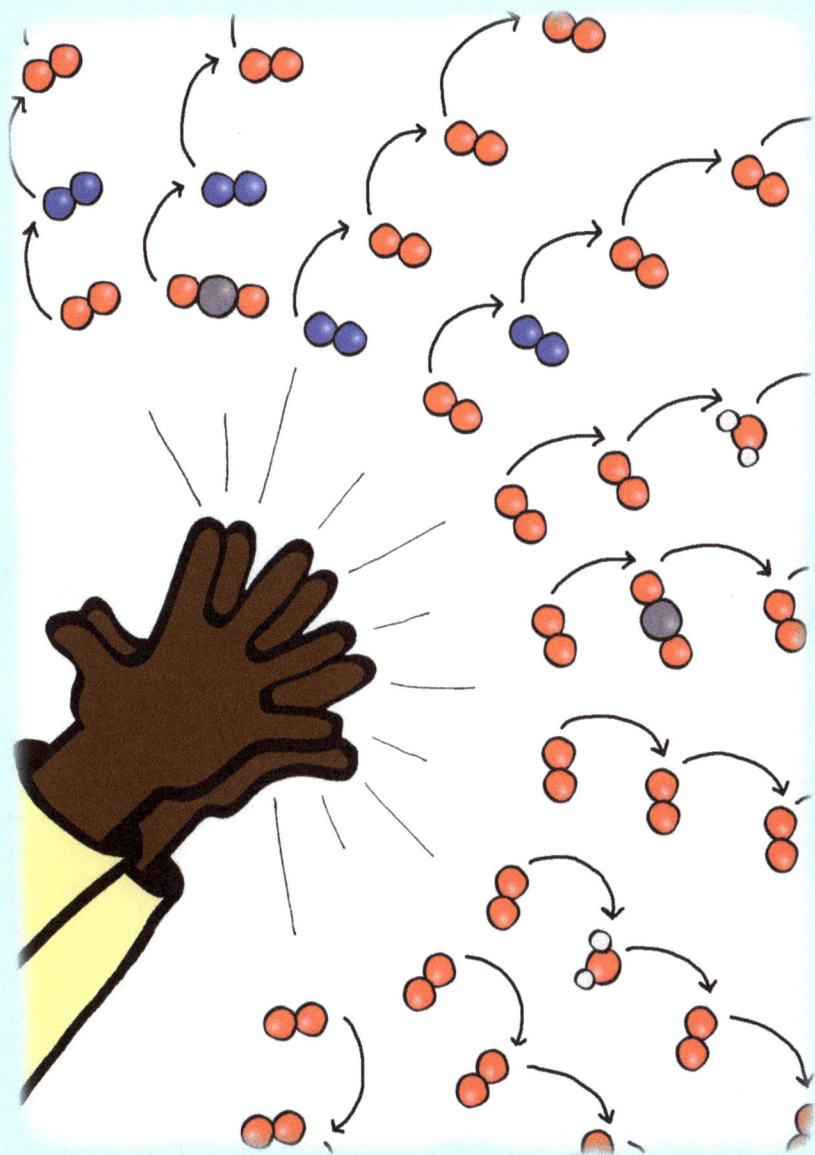

Molecules are made when atoms link together.

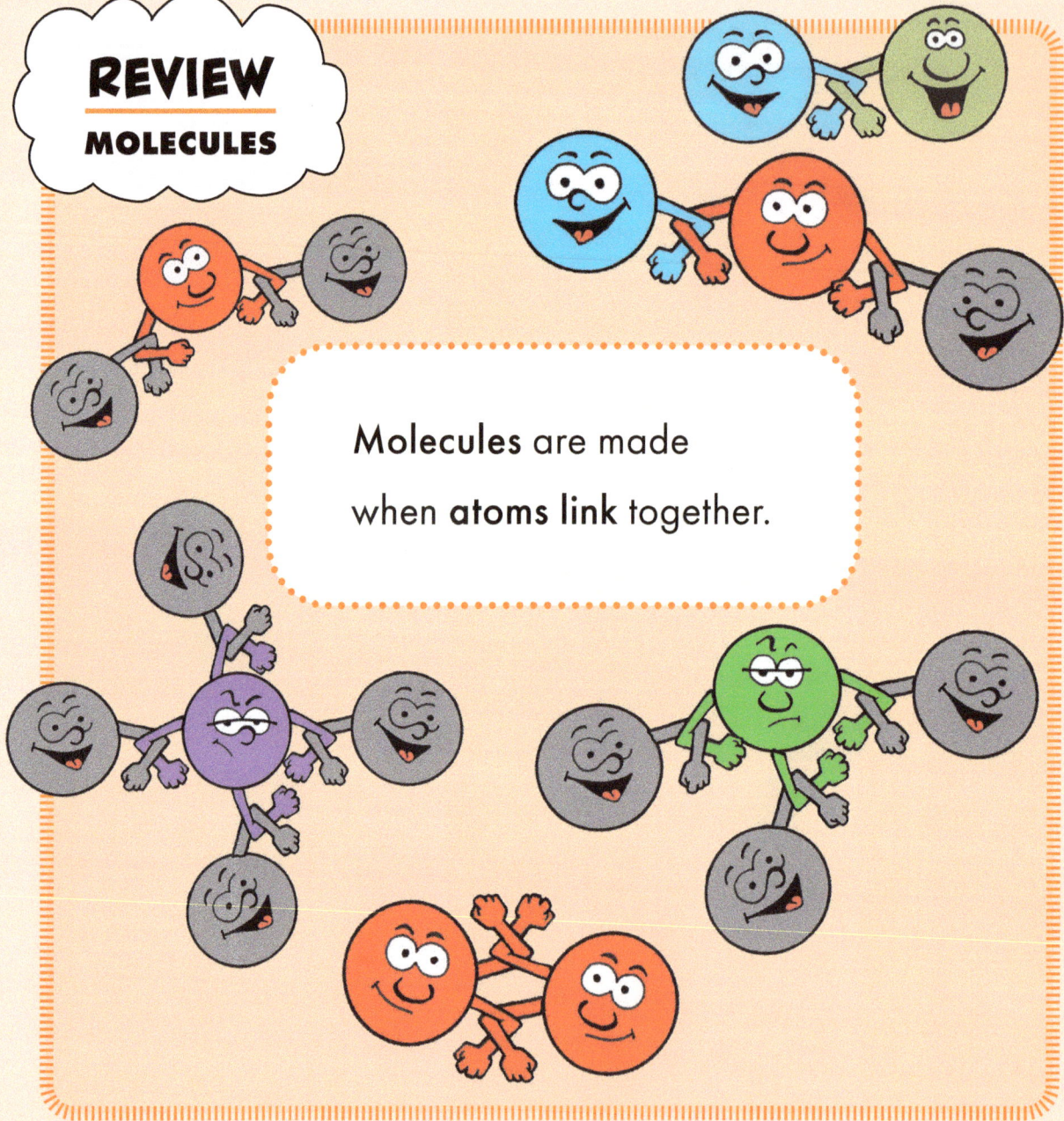

Atoms are tiny building blocks that can link together.

Atoms make up everything we touch, taste, smell, and see.

As the air molecules wiggle around, the energy of the wiggling movement is passed along from molecule to molecule.

Energy is passed from
molecule to molecule.

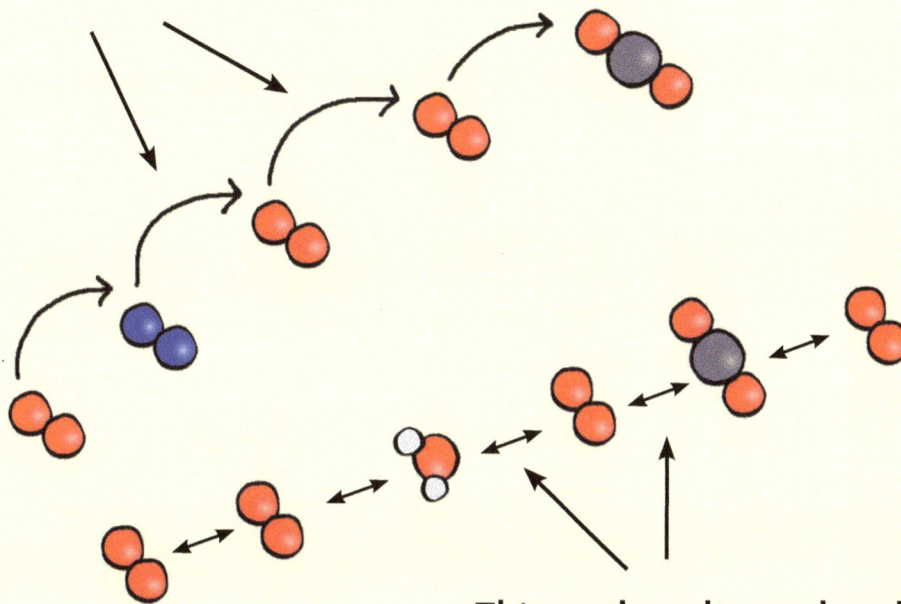

This makes the molecules
wiggle slightly back and forth.

Energy is needed to do **work.**

Work happens when a **force** moves an object.

Force is any action that changes...

...the **location** of an object,

...the **shape** of an object,

...**how fast or how slowly** an object is moving. (This is called the **speed** of an object.)

This passing of energy in a sound wave creates a longitudinal wave.

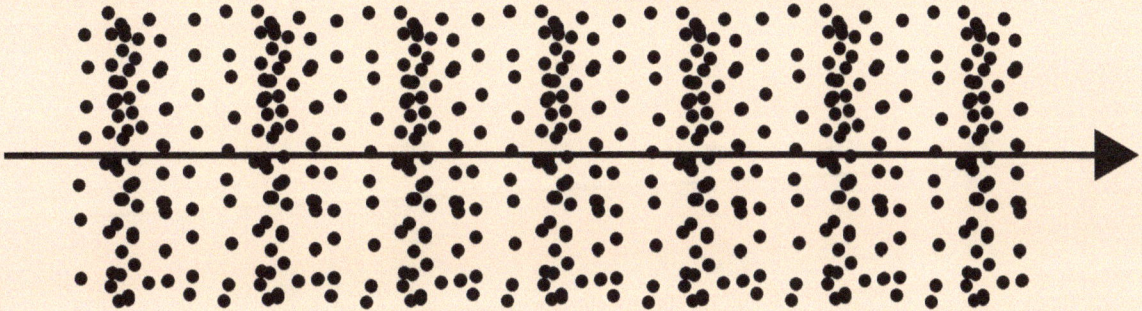

Molecules in a Longitudinal Wave

A **longitudinal wave** occurs when **energy** is transferred to surrounding **molecules**.

The molecules in a longitudinal wave move slightly back and forth.

Molecules before a wave passes through them.

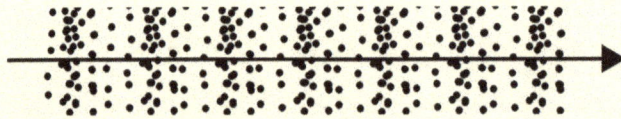

Molecules as a wave passes through them.

The more energy that is used, the bigger the longitudinal wave and the louder the sound.

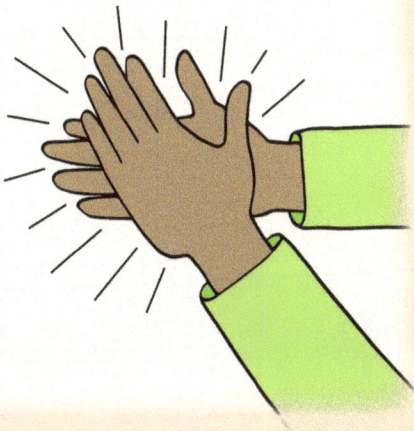

Try it. Clap softly then hard.

Can you tell the difference?

Your ear picks up sound with a thin flexible part called the **tympanic membrane.** The tympanic membrane picks up the motion of the molecules in the sound wave that is traveling through the air.

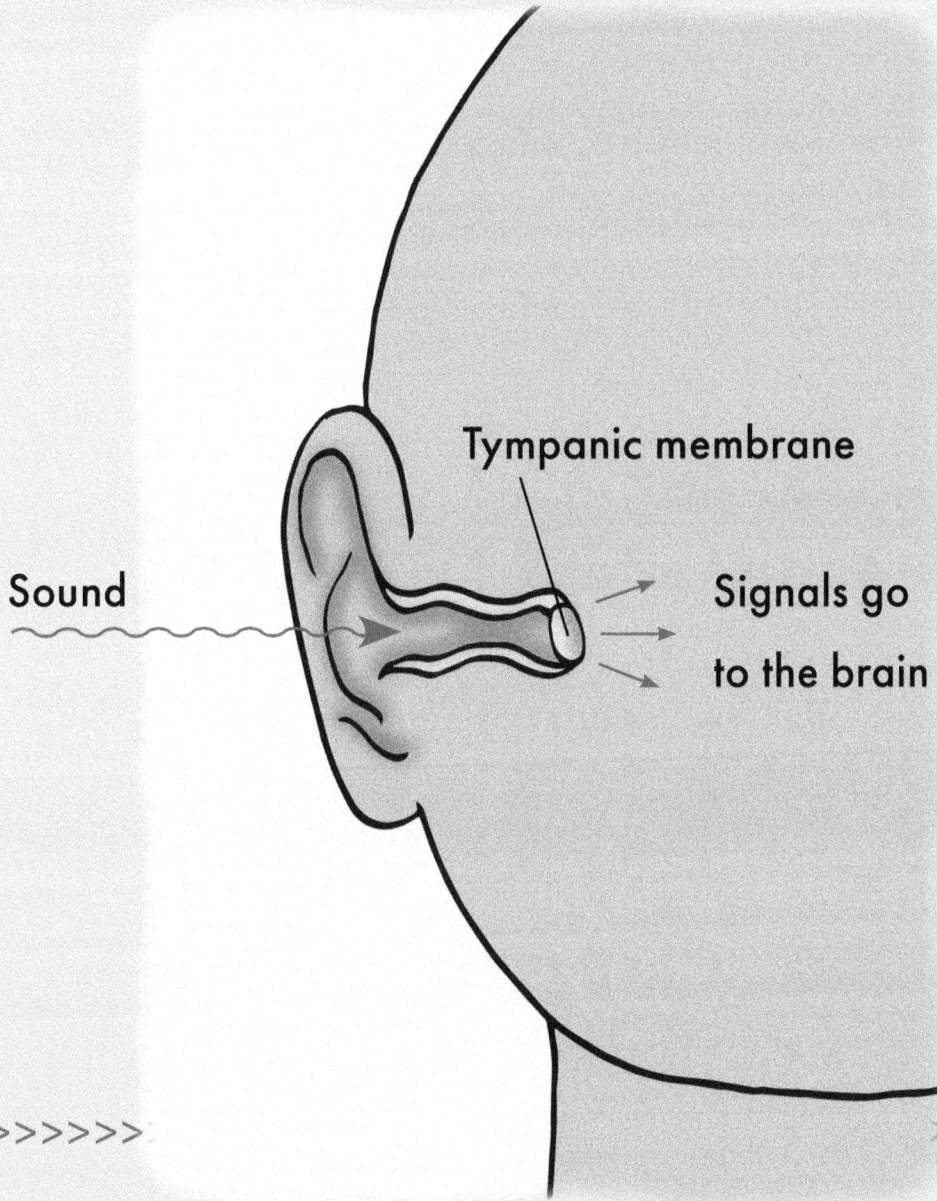

Sound

Tympanic membrane

Signals go
to the brain

A sound wave can travel through other materials besides air.

Try this little experiment.
Put your ear on a desk or table.
Tap the desk or table with your knuckles.

Can you hear the sound traveling through the wood?

The energy from the tapping wiggles the wood molecules. This creates a wave that your ear picks up as sound.

Wood molecules

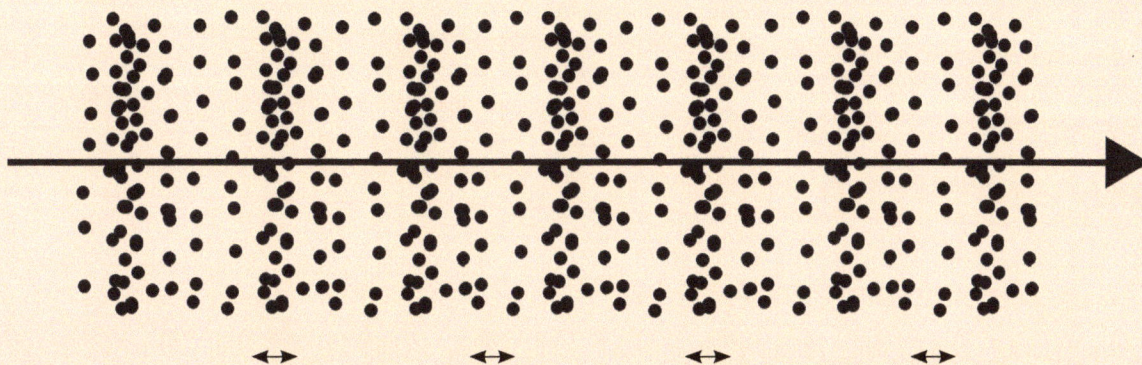

When the wood molecules wiggle,
they move side to side slightly.
They get a tiny bit closer together
and a tiny bit farther apart.

Sounds travel faster in water and in metals than in air because the molecules are more tightly packed in water and metals. This means energy can be more easily transferred from molecule to molecule.

Hello!

Hi!

The Speed of Sound in Different Materials

(Meters per Second)

Air	343
Water	1497
Aluminum	6320

You can guess from this chart that air has molecules that are farthest apart. Aluminum has molecules that are closest together.

How to say science words

atom   (AA-tum)

energy   (E-nuhr-jee)

location   (loh-KAY-shun)

longitudinal   (lawn-juh-TOOD-nuhl)

molecule   (MAH-lih-kyool)

force   (FAWRSS)

shape   (SHAYP)

speed   (SPEED)

tympanic membrane   (tim-PAA-nik MEM-brayn)

wave   (WAYV)

work   (WERK)

What questions do you have about SOUND?

Learn More Real Science!

Complete science curricula from Real Science-4-Kids

Focus On Series

Unit study for elementary and middle school levels

Chemistry
Biology
Physics
Geology
Astronomy

Exploring Science Series

Graded series for levels K–8. Each book contains 4 chapters of:

Chemistry
Biology
Physics
Geology
Astronomy

www.ingramcontent.com/pod-product-compliance
Lightning Source LLC
Chambersburg PA
CBHW040152200326
41520CB00028B/7579